Victoria Vasilyevna Savelyeva

Non-contact monitoring technologies for electrochemical reactions

Victoria Vasilyevna Savelyeva

Non-contact monitoring technologies for electrochemical reactions

Practical application of technology for monitoring and evaluation of complex processes

ScienciaScripts

Imprint

Cover image: www.ingimage.com

This book is a translation from the original published under ISBN 978-620-7-45415-0.

Publisher:
Sciencia Scripts
is a trademark of
Dodo Books Indian Ocean Ltd. and OmniScriptum S.R.L publishing group

120 High Road, East Finchley, London, N2 9ED, United Kingdom
Str. Armeneasca 28/1, office 1, Chisinau MD-2012, Republic of Moldova, Europe
Printed at: see last page
ISBN: 978-620-8-29515-8

Contents

Maintaining

People perceive information through photos and videos with attached description, history of product creation. It requires a professional approach and a special methodology for photo documentation of processes to make analytics. Step-by-step fixation of the object by a specially developed technology, which I am the author of, will give a full perception of the development, as well as give control from the side of investors and the management group of the product development project. Let us take the example of the invention of a capsule for monitoring the acidity of bovine gastric juice with an inbuilt electrochemical battery. One of the techniques of dimensional representation of a product is also given. The dimensional representation of a product is an important aspect in the design, manufacturing and transport process. It involves determining the physical dimensions of the product, which may include length, width, height, and weight. Here are a few techniques for dimensional representation of a product:

1. CAD (Computer-Aided Design): The use of computer-aided design allows the creation of accurate 3 D models of a product that can be used to determine its overall dimensions. These models can be scaled, rotated and viewed from different angles to obtain accurate dimensional information.
2. Physical prototypes: Creating a physical prototype of a product allows you to get a realistic idea of its overall dimensions. This can be particularly useful when designing large or complex products.
3. Use of standards: There are various standards and guidelines that define how product dimensions should be measured and presented. This can include everything from generally accepted international standards to specific industry guidelines.
4. Virtual Reality: Using VR technology, designers and engineers can visualise and measure the dimensions of a product in a virtual

environment. This allows for a more accurate visualisation of how the product will look and function in the real world.

5. Use of mathematical models: For some products, especially those with complex shapes or dimensions, it may be useful to use mathematical models to determine overall dimensions. This may involve the use of geometry, trigonometry or other mathematical methods.

Depending on the product and its requirements, one or more of these techniques may be used. The key is to provide an accurate and reliable dimensional representation of the product to ensure an efficient design, manufacturing and transport process.

Recently there is a lot of discussion among professionals and society about innovative products, because with the development of social networks and technological boom there is a huge flow of information, the delivery of information from the developer to the consumer, has become an instant. People perceive information through photos and videos with attached description, history of product creation.

Thus, the main thing in the creation of the product was a system of graphic and photographic support of the process at all stages.

A graphic and photographic process support system is a key element in project and process management. It is a visualisation of the different stages of a process, which helps to better understand its progress and control tasks.

1. Graphical support: This may involve creating charts, graphs and diagrams that show the different stages of the process and their interrelationships. For example, a Gantt chart is used to display a planned work schedule, showing the start, duration and completion of each task. A Perto diagram allows you to visualise the dependencies between tasks and identify the critical path of a project.

2. Photo accompaniment: This includes the use of photographs to documenting the process. Photographs can be used to track work progress, identify problems, or simply to document the work performed. They can be particularly useful in processes where visual information is important, such as construction or manufacturing.

3. Integrating graphics and photos: Using these two methods together can give the most complete picture of the process. Graphical diagrams and charts can show the overall structure and flow of the process, while photographs can provide a detailed view of specific stages of the work.

4. Use of technology: Modern technology, such as project management software, can automate the process of creating graphics and photo accompaniments. This can simplify the management process and make it more efficient.

As in any business, when a product is created by professionals, large teams and corporations, it requires a professional approach and a special methodology for photo-documentation of processes to make analytics, to help developers and to control processes, and subsequently production, in its correct presentation, for the consumer of the product. Such a system will be discussed in this book.

Herein is one example of the invention of a capsule for monitoring the acidity of bovine gastric juice with an integrated electrochemical battery, showing circuits and devices by 3D construction in computer programs and step-by-step recording of results using a specially designed programme developed by me.

The result was to open the curtain of the product development, how it looks like and how the tests took place, and to describe such tools as the system of graphic photo support of the innovative product development process with the use of a special programme developed by me.

The impact of my specially designed programme to graphically support the product creation process has had great results, which are presented here. We know that when creating any invention, the development takes place in 3D programmes on computers. But in most cases, at all stages of the production of an invention, it is necessary to be able to properly conduct graphic and photographic support of the process. I have developed a special programme of graphical photo-support of the process, where a special step-by-step technique of photofixation of the product reproduction is used. This programme works effectively for any innovative product. There is a wonderful expression that it is better to see once than to hear 100 times. The human perception is so arranged that the object he sees in person is already in itself an indisputable perception of information about it, namely size, colour and its dimensions. When a person watches a presentation or a 3D model he cannot fully and completely visualise the future invention in its entirety and keep it in his memory. This decision affects the speed of development and decision-making on the product at all stages. Now, in the modern world, the speed of bringing the product to the market is important, and most importantly, the history of development and subsequent presentation of the object to have commercial success. Step-by-step fixation of the object according to the specially developed technology of Victoria Savelieva, where angles, use of light and focal length are taken into account, will give a full perception of the development, as well as give control from the side of investors and the management group of the product development project. Also, special technology is used to photograph directly the project executors at their workplace and the decision-making group and focus group for product evaluation, where it is possible to trace the psychological state of the body during the development. All this is included in the analytical support of the product, where it is possible at all stages to track the overall dynamics of the development and make adjustments that will affect the output of the product.

Often a person is controlled by his emotions. Let's consider an individual person in the team who was able to make a certain result on some part of the development and his correct, according to special technology of photo-fixation, emotions at the moment of obtaining the result, at the subsequent briefing of broadcasting these emotions on the photo, can affect the whole team, to reveal the other participants' potential and healthy competition. Why is this necessary? - Let's consider the example of the invention of a capsule for monitoring the acidity of bovine gastric juice with an integrated electrochemical battery, which is given below. The author wants to ask the audience of this research paper: "how much do you visualise this invention in person and how much would you be influenced by a photo of the product 'as seen in person' to understand this invention and its application when it has been filmed from its inception and the stages of testing and application practice".

CHAPTER 1

PRELIMINARY TECHNICAL PROPOSAL OF THE FEASIBILITY STUDY PROJECT - "CAPSULE FOR MONITORING BOVINE GASTRIC JUICE ACIDITY WITH INTEGRATED ELECTROCHEMICAL BATTERY"

System and method of fuel cells for electricity storage

The invention discloses an innovative fuel cell type electrochemical energy conversion device and a method of potentially high power density, suitable in particular for autonomous applications such as electrically driven vehicles.A converted rechargeable battery comprising plates, electrolyte and separators which can be similar to a conventional known lead-acid battery. Thus, it can be charged from an electrical power source or discharged from an external load with the typically high surge current characteristics of rechargeable batteries. However, the main purpose of the present invention is to provide an innovative mode of additional charging of the fuel cell-storage device by chemical reactions of each electrode with fuel and oxidant supplied at specific flow ratios. While all types of prior art fuel cells exchange ions between electrodes via an electrolyte, each of the electrodes of the present fuel cell independently exchanges ions using only an adjacent electrolyte.

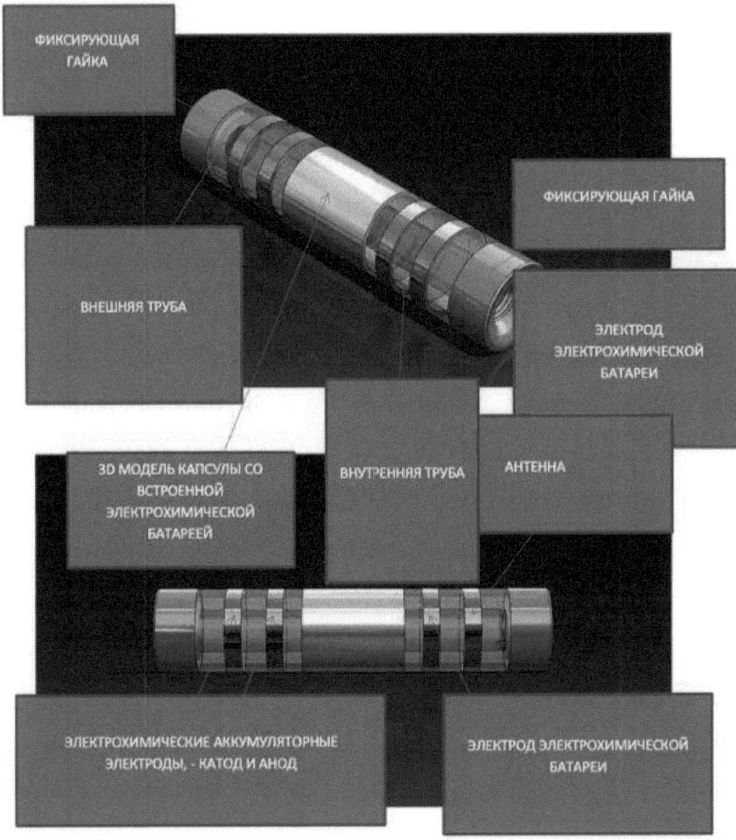

FIXING **NUT** ELECTROCHEMICAL BATTERIES **ELECTRODES, -** BATTERY **CATHODE AND ANODE** **ELECTRODE** **ELECTROCHEMICAL** BATTERIES	**3D MODEL OF THE CAPSULE WITH** **INTRODUCED** **ELECTROCHEMICAL** **BATTERY** **RETAINING NUT** EXTERNAL PIPE INNER TUBE ANTENNA

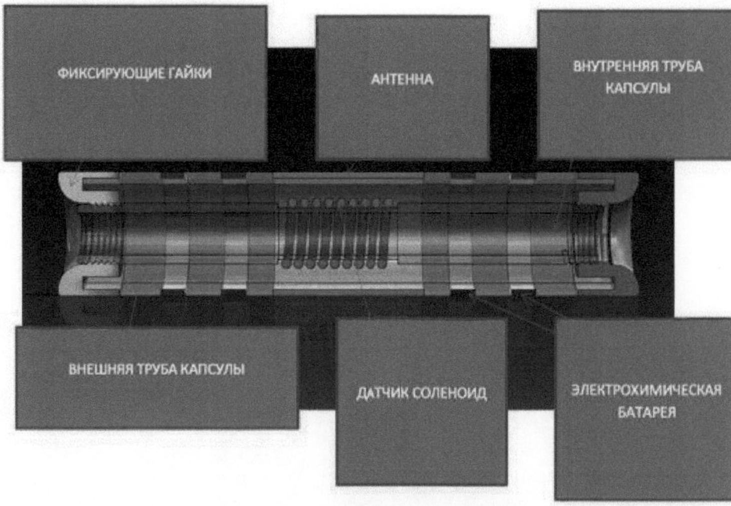

FIXING NUTS ANTENNA INNER TUBE

EXTERNAL CAPSULE TUBE SENSOR SOLENOID SOLENOID ELECTROZYMIC BATTERY

Figure 1. External ZD view of the capsule model

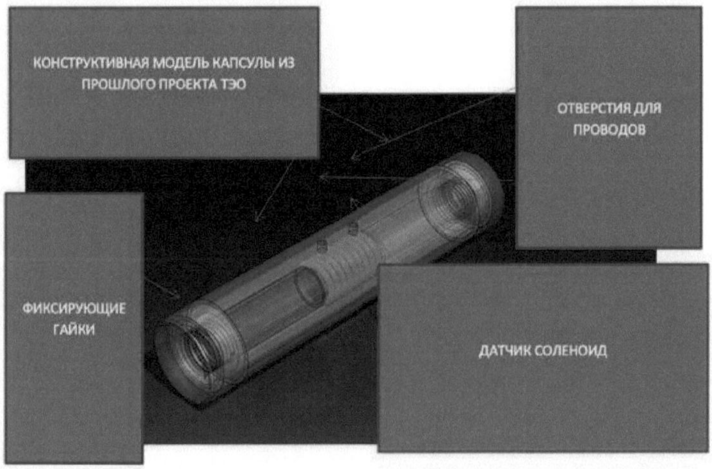

CROSS SECTION OF THE PROPOSED CAPSULE DESIGN WITH INTEGRATED ELECTROCHEMICAL BATTERIES

STRUCTURAL MODEL OF THE CAPSULE FROM THE PAST FEASIBILITY STUDY PROJECT
WIRE OPENINGS
LOCKING NUTS SENSOR SOLENOID

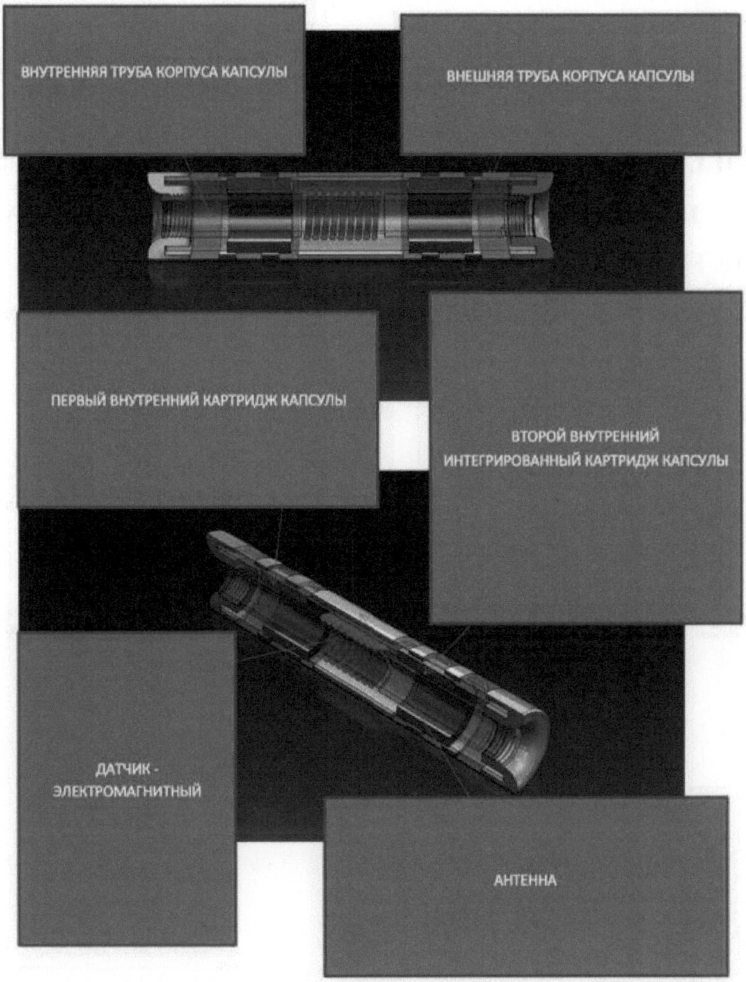

INNER TUBE OF THE CAPSULE BODY OUTER TUBE OF THE CAPSULE BODY

FIRST INTERNAL CAPSULE CARTRIDGE SECOND INTERNAL INTEGRATED CAPSULE CARTRIDGE SECOND INTERNAL INTEGRATED CAPSULE CARTRIDGE SECOND INTERNAL INTEGRATED CAPSULE CARTRIDGE

ELECTROMAGNETIC SENSOR ANTENNA

CHAPTER 2

ACIDIC COW STOMACH IS THE EQUIVALENT OF AN ELECTROLYTE

THE FOLLOWING PATENTS PROVIDE EXAMPLES OF THE USE OF THE IDEA OF PRODUCING THE ODES NECESSARY FOR THE FUNCTIONING OF AN INTEGRAL CAPSULE.

OF THE ELECTROCHEMICAL BATTERY:

Electrolyte for galvanic cell battery

An electrolyte for an electrochemical battery cell containing sulphur dioxide and a conductive salt. The electrolyte is a gel formed with fluorosulfinate. The invention is also directed to a battery cell which contains such an electrolyte.

The electrolyte in an electrochemical battery cell plays a key role in the charge transfer process between the anode and cathode. It provides the conductivity of ions that are necessary for the chemical reaction process that takes place in the battery.

1. What is an electrolyte?

An electrolyte is a medium or substance that has free ions capable of conducting an electric current. In the context of batteries, electrolytes are usually categorised as solid, liquid or gel substances.

2. The role of the electrolyte in the battery

In a battery, the electrolyte serves as a bridge for ion exchange between the anode and cathode. When the battery is charged or discharged, ions move through the electrolyte from one electrode to the other, causing a chemical reaction that results in the flow of electric current.

3. Types of electrolytes

Liquid electrolytes are the most common type of electrolytes used in batteries. They usually consist of a solution of salts, acids or bases. However, they have some disadvantages such as the possibility of leakage or evaporation.

Solid electrolytes are materials, usually salts, that can conduct ions in the solid state. They offer greater safety because they are less likely to leak or explode.

Gel electrolytes represent a compromise between liquid and solid electrolytes. They have the properties of both liquid and solid electrolytes, providing good ion conductivity and safety.

4. Electrolyte selection

The choice of electrolyte for an electrochemical battery cell depends on many factors, including required battery performance, application, cost and safety. For example, high-energy applications such as electric vehicles may require electrolytes with high ion conductivity.

In general, the electrolyte is a key component in an electrochemical battery cell, playing an important role in the charge transfer process and ensuring battery performance.

Integral electronic device and methods of its manufacture

An integrated electronic device and a method of manufacturing thereof are provided. The integrated electronic device may include an electronic assembly, such as an RFAID active assembly, which is electrically coupled to a thin printed flexible galvanic cell. In one example, the electronic assembly and the electrochemical battery are located on the same substrate. In one exemplary manufacturing method, the entire cell must be fabricated on a printing press to integrate the battery directly into the electronic unit.

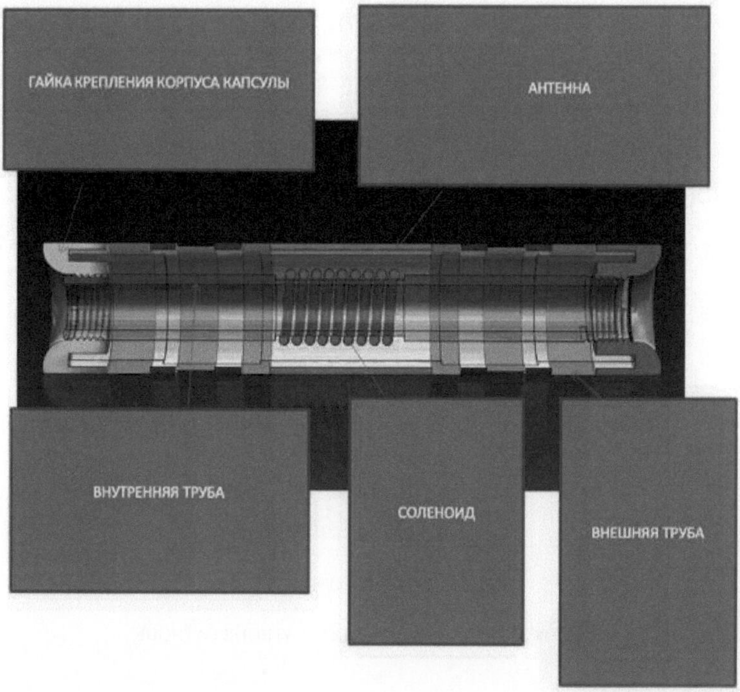

CAPSULE BODY ATTACHMENT NUT ANTENNA

INNER TUBE SOLENOID OUTER TUBE

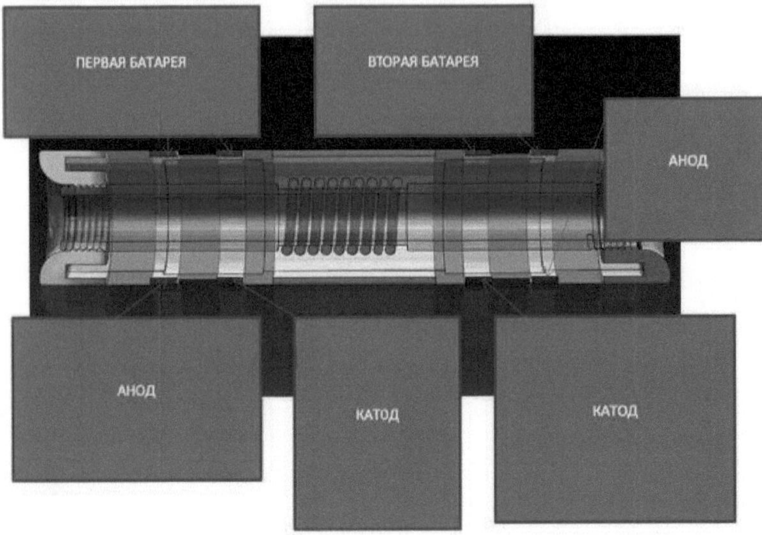

FIRST BATTERY SECOND BATTERY

ANODE ANODE ANODE CATHODE CATHODE CATHODE

CHAPTER 3

Programme excerpt.

Dimensional representation of the product.

One of the important points of photographic fixation is to photograph the invention with objects which any person has at hand or constantly encounters with these objects. A person can easily visualise the dimensions of the invention and then he will have a fixation of the volumes in his memory, and later, when describing its use and other tests, the processes will be clear. It also simplifies the communication of information on complex parts of the development.

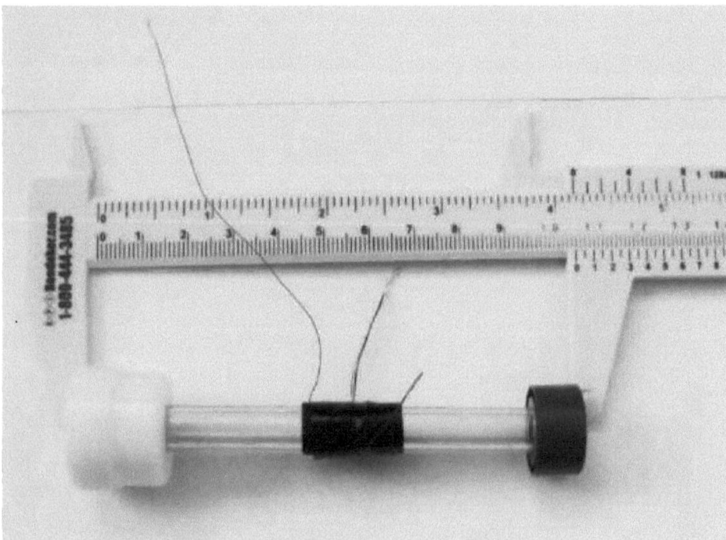

Figure 6. Dimensioning in the design process

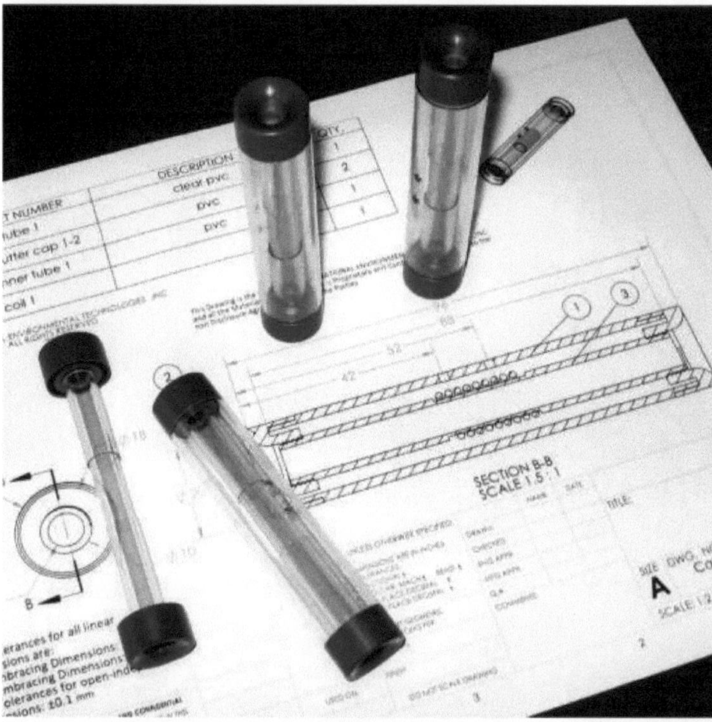

Figure 7. Snapshot showing capsule dimensions view 1

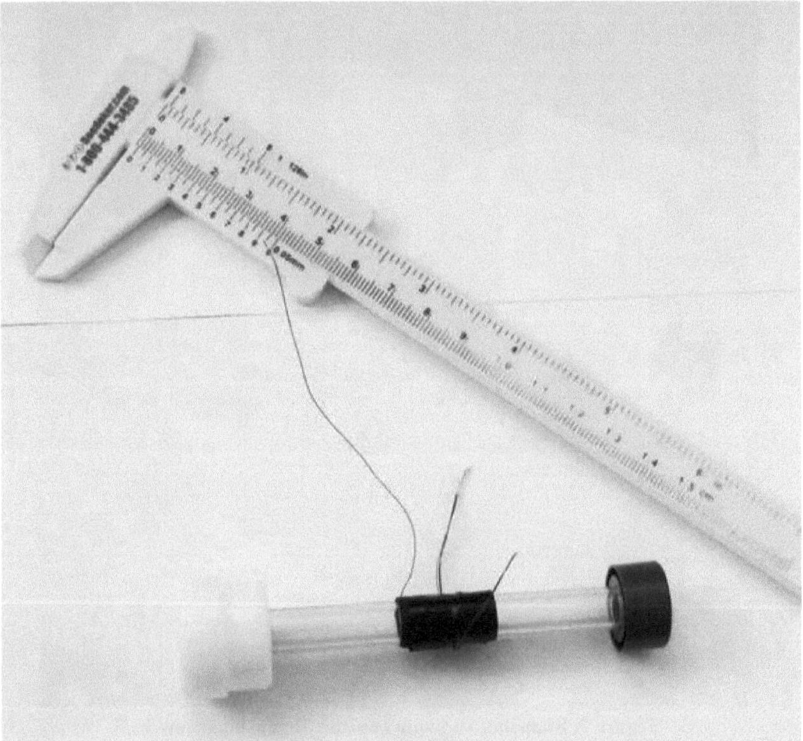

Figure 8. Snapshot showing capsule dimensions view 2

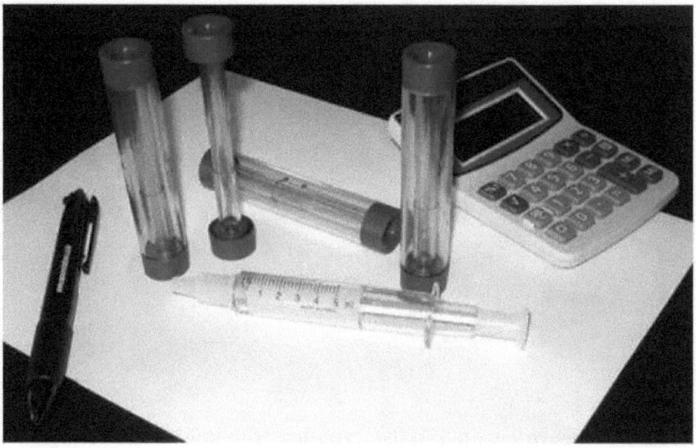

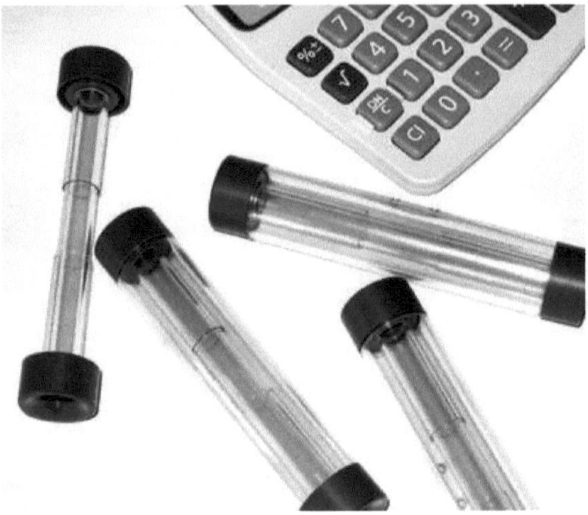

Figure 9. Comparative exposure with items view 1

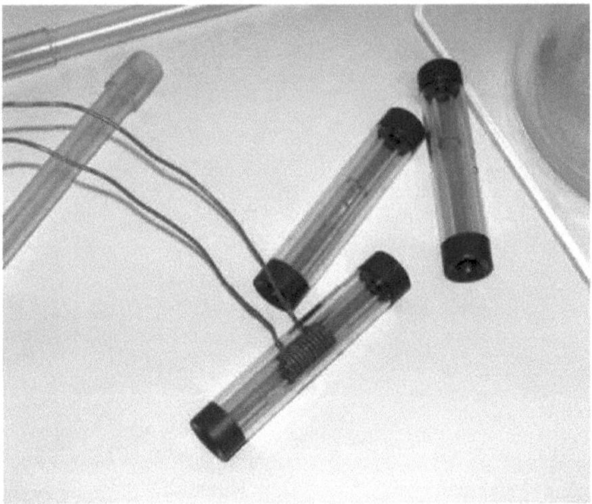

Figure 10. Test papers

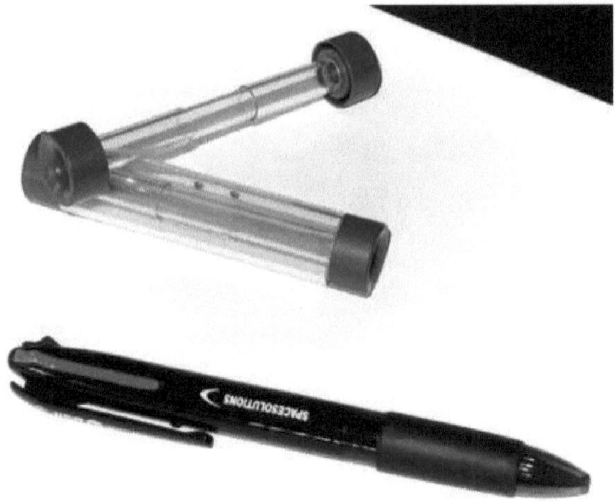

Figure 11. Comparative exposure with objects view 2

List of references

PRELIMINARY TECHNICAL PROPOSAL FOR A DRAFT FEASIBILITY STUDY ON THE TRIAL PROJECT - "CAPSULE FOR MONITORING THE ACIDITY OF BOVINE GASTRIC JUICE WITH INTEGRATED ELECTROCHEMICAL BATTERY"

PART 1

SEARCH FOR PATENTS AND PATENT APPLICATIONS

IN THIS TECHNOLOGICAL FIELD WAS FOUNDED AFTER PATENTS PARTLY RELATED TO THE PROPOSED SUBJECT MATTER WERE OBTAINED:

U.S. Patent

Uchiyama , et al.

8,530,112

10 September 2013.

A device for generating electricity

Annotation

A device for generating electricity includes: a tubular fuel cell having an electrolyte layer disposed between an inner electrode and an outer electrode to which a fuel gas is supplied, the fuel cell having an inner portion formed as an inner channel for the fuel gas, a protective tube arranged around the fuel cell with a gap provided between the outer electrode and the protective tube, a connecting element connecting the fuel cell and the protective tube to each other and allowing the outer channel to be formed for the

Inventors: Uchiyama; Naoki (Hamamatsu, JP), Uchiyama; Yasuyuki (Hamamatsu, JP), Nakabayashi; Seigou (Hamamatsu, JP)

Applicant: NameCityStateCountryType

Uchiyama; Naoki Hamamatsu n/a JP

Uchiyama; Yasuyuki Hamamatsu N/A JP

Nakabayashi; Seigou Hamamatsu N/A JP

Recipient: Kabushiki Kaisha Atsumitec (Shizuoka, JP)

Family ID :44762344

Application No.: 13/640,197

Filed: 21 February 2011.

PCT Submitted: 21 February 2011.

PCT Number: PCT/JP2011/053679.

371(c)(1),(2),(4)

09 October 2012.

Date:

PCT Patent. Number: WO2011/125378

PCT Pub. Date: 13 October 2011.

U.S. Patent 8,496,791

Takeuchi et al. 30 July 2013.

Active oxygen apparatus

Annotation

A device for producing active oxygen includes cathodes comprising a plurality of base materials each containing a conductive polymer, an anode having conductivity, a power source conducting electricity between both electrodes through water in which oxygen is dissolved, and a water receiving portion containing water. The cathodes comprise a plurality of plate-like base materials mounted vertically at spaced intervals in the water receiving portion, and the anode is arranged across the plurality of base materials and orthogonal to the plurality of base materials.

Inventors: Takeuchi; Shiro (Chiyoda-ku, JP), Saito; Mari (Chiyoda-ku, JP), Furuhashi; Takuya (Chiyoda-ku, JP)

Applicant: Name City State Country Type

Takeuchi; Shiro Chiyoda-ku N/AJP

Saito; MariChiyoda-kun/AJP.

Furuhashi; Takuya Chiyoda-ku N/AJP

Consignee: Mitsubishi Electric Corporation (Chiyoda-ku, Tokyo, JP)

Family ID :42982212

Application No.: 13/059,668

Filed: 15 April 2009.

PCT Filed: 15 April 2009.

PCT Number: PCT/JP2009/057594.

371(c)(1),(2),(4)

Date: 18 February 2011.

PCT Patent. Number: WO2010/119529

PCT Pub. Date: 21 October 2010.

U.S. Patent8 ,404,099

Fowler

26 March 2013.

Electrolysis of spent fuel pool water to produce hydrogen

Annotation

A method of electrolytic hydrogen production in which radiation-excited water from the spent fuel pool of a nuclear power plant is fed into one or more electrolysers, where direct current is applied to pairs of electrodes in the electrolysers to form hydrogen and oxygen. The hydrogen is collected. Hydrogen can be collected in an energy storage system to produce large amounts of hydrogen when electricity demand is low and little or no hydrogen when electricity demand is high.

Inventors: Fowler; David E. (Gainesville, Fla.)

Applicant: NameCityState Country Type

Fowler, David E. Gainesville, Florida, USA

Family ID: 42036519

Application No.: 12/552,469

Filed: 2 September 2009.

U.S. Patent 8,354,195

Brandstetter et al. 15 January 2013.

Electric storage fuel cell system and method

Annotation

The invention discloses an innovative device and method for electrochemical energy conversion of the fuel cell type with potentially high power density, suitable in particular for autonomous applications such as vehicles powered by electricity The inventive device, called "fuel cell", can be considered as a converted rechargeable battery comprising plates, electrolyte and separators, which can be similar to a conventional known lead-acid battery. Thus, it can be charged from an electrical power source or discharged through an external load with the typically high surge currents typical of rechargeable batteries. However, the main object of the present invention is to provide an innovative additional charging mode for the rechargeable fuel cell by chemically reacting each electrode with fuel and oxidant supplied at specific flow ratios. While all types of prior art fuel cells exchange ions between electrodes via an electrolyte, each of the electrodes of the present fuel cell exchanges ions independently using only its adjacent electrolyte.

Inventors: Brandstetter; Aharon (Kfar-Vradim, IL), Brandstetter; Haim (Bnei-Beraq,

IL)

Applicant: NameCityStateCountryType

Brandstetter, Aaron Kfar-Vradim, Ill.

Brandstetter; Haim Bnei-Berak N/A IL

Family ID: 40379104

Application No.: 12/344,964

Submitted: 29 December 2008.

U.S. Patent 8,295,031

Lee et al. 23 October 2012.

Electrical double layer capacitor and its fabrication method

Annotation

An electrical double layer capacitor cell is provided, comprising: an outer casing having a housing space formed of an insulating resin; first and second outer leads recessed in the outer casing, each having a first surface extending into the housing space and a second surface extending to an outer side of the outer casing; and a chip type electrical double layer capacitor cell disposed in the housing space and electrically coupled to the first surface. The electrical double layer capacitor cell includes first and second electrodes facing each other and having electricity of opposite polarities, at least one induction electrode layer disposed between the first and second electrodes and having no electricity applied thereto, and first and second separators disposed between the first electrode and the induction electrode layer and between the second electrode and the induction electrode layer, respectively.

Inventors: Lee; Sang Kyung (Kyunggi-do, ROC), Park; Il Kyu (Seoul, ROC), Han; Seung

Hong (Seoul, KR), Noh; Jung Eun (Gyeonggi-do, KR)

Recipient: Samsung Electro-Mechanics Co., Ltd. (Suwon, ROC)

Family ID: 44258363

Application No.: 12/805,761

Filed: 18 August 2010.

U.S. Patent8 ,178,255

Nishikawa et al. 15 May 2012.

Fuel cell

Annotation

The fuel cell of the present invention generates electricity by supplying a fuel fluid to one of a pair of electrodes forming the MEA 1 and supplying an oxidising fluid to the other electrode, wherein at least one of the fuel and oxidising fluid is a gas, includes a gas supply device moving the gas along a flow path 10 defined on the surface of the MEA 1, and a drive circuit actuating the gas supply device including a vibrating plate 4 and a reflective wall on either side of the flow path 10, and the drive circuit performs normal operation control by generating a flow of gas from an inlet to an outlet of the flow path 10 due to a sound pressure gradient generated in the flow path 10 by vibration of the vibrating plate 4 and the foreign wall

The material elimination operation eliminates foreign material in the flow path 10 by changing the vibration mode of the vibratory plate 4.

Inventors: Nishikawa; Masato (Moriguchi, JP), Nakagawa; Tatsuyuki (Moriguchi, JP), Kihara; Hitoshi (Moriguchi, JP).

Consignee: SANYO Electric Co., Ltd. (Moriguchi, JP)

Family ID :39838174

Application No.: 12/525,698

Filed: 31 January 2008.

PCT Filed: 31 January 2008.

PCT Number: PCT/JP2008/051495.

371(c)(1),(2),(4)

Date: 04 August 2009.

PCT Patent. №: WO2008/096657.

PCT Pub. Date: 14 August 2008.

U.S. Patent8 ,158,277

Fischel

17 April 2012.

Electrochemical batteries with cross-flow

Annotation

A cross-flow electrochemical cell for generating electricity is disclosed, which includes means for cross-pumping electrolyte through anode and cathode electrodes in the same direction to achieve substantially higher discharge and charge currents. The cross-pumping enables the use of thick mesh electrodes comprising scaffolds impregnated with metal nanoparticles having high surface area and high porosity.

Inventors: Fischel; Halbert (Santa Barbara, Calif.)

Recipient: Global Energy Science, LLC (California) (Santa Barbara, California)

Family ID: 45694441

Application No.: 13/171,080

Filed: 28 June 2011.

U.S. Patent8 ,105,722

Leonida 31 January 2012.

A fuel cell system suitable for using fossil fuels and its method of operation

Annotation

The fuel cell system includes a first electrolyte electrode, a first electrode having a first electrode coupled to one side of the first electrolyte electrode and a second electrode coupled to an opposite side of the first electrolyte electrode, and a second electrolyte electrode having a third electrode coupled to one side of the second electrolyte electrode and a fourth electrode coupled to an opposite side of the second electrolyte electrode. The first channel is in fluid communication with the first electrode, and the second channel is in fluid communication with the fourth electrode and an electrically conductive grid disposed between the second electrode and the third electrode. Portions of the second and third electrodes are coupled to each other through openings defined by the mesh. The fuel cell system further includes an electrical power source connected to the first and second electrodes; and an electrical circuit connected to the first and second channels.

Inventors: Leonida; Andrew (West Hartford, CT)

Family ID: 44656870

Application No.: 13/154,886

Filed: 7 June 2011.

Ruhani

22 October 1996

A method and apparatus for measuring the state of charge of a battery based on the volume of battery components

Annotation

The state of charge of electrochemical batteries of various kinds is determined by measuring the gradual change in the total volume of reactive masses in the battery. The invention is based on the principle that all electrochemical batteries, both primary and secondary (rechargeable), generate electricity by chemical reaction with at least one electrode, and that the chemical reactions result in certain changes in the composition and density of the electrode. The electrode reaction mass, the electrolyte, and any separators or gaskets are typically contained within a battery case of some volume. As the battery is used or recharged, the specific volume of at least one of the electrodes will change, and since the masses of materials do not change significantly, the total volume occupied by at least one of the electrodes will change. These volume changes can be measured in a variety of ways and are related to the state of charge of the battery. In one embodiment, the change in volume can be measured by observing small changes in one of the major dimensions of the battery case as it expands or contracts to accommodate the total volumes of its components.

Inventors: Rouhani; S. Zia (Idaho Falls, ID).

Recipient: Lockheed Idaho Technologies Company (Idaho Falls, ID)

Family ID: 23612635

Application No.: 08/407,570

Filed: 21 March 1995

Current US class: 429/93 ; 429/61; 429/66; 429/70; 429/90; 429/91

Current CPC class: H01M 6/5044 (20130101); H01M

10/48 (20130101); Y02E 60/12 (20130101)

Current InternationalH01M 10/48 (20060101); H01M 10/42 (20060101);

Class: H01M 6/00 (20060101); H01M 6/50 (20060101);

H01M 010/48 ()

Search area:

U.S. Patent8,599,572

Neudecker , et al.

3 December 2013.

Printed circuit board with integrated thin-film accumulator

Annotation

The present invention relates, for example, to printed circuit boards having a thin film battery or other electrochemical element between or within its layer or layers. The present invention also relates, for example, to electrochemical elements arranged in a stack of layers of a printed circuit board.

Inventors: Neudecker; Bernd J. (Littleton, CO), Keating; Joseph A. (Broomfield, CO)

Applicant: NameCityState Country Type

Neudecker, Bernd J. Littleton, USA

Keating, Joseph A. Brumfield, USA

Receiver: Infinite Power Solutions, Inc. (Littleton, CO)

Family ID: 43623158

Application Number: 12/873,953

Filed: 1 September 2010.

U.S. Patent 8,574,762

Dan, et al. 5 November 2013.

Metal-oxide negative electrodes for lithium-ion electrochemical cells and batteries

Annotation

Compositions of negative electrodes for lithium-ion electrochemical cells including metal oxides and polymer binders are presented. Also presented are electrochemical cells and battery packs incorporating electrodes made using these compositions.

Inventors: Dahn; Jeffrey R. (Halifax, CA), Li; Jing (Halifax, CA), Obrovac; Mark N. (St. Paul, MN)

Applicant: NameCityState Country Type

Dun; Jeffrey R. Halifax N/A California

Lee; JingHalifax n/a California

Obrowatz, Mark N. St. Paul, New Mexico, USA

Recipient: 3M Innovative Properties Company (St. Paul, Massachusetts)

Family ID: 41131621

Application No.: 13/224,925

Filed: 2 September 2011.

U.S. Patent 8,518,578

Swan 27 August 2013.

**Please see images: (Certificate of Correction) **

Electrode plate for electromechanical battery

Annotation

An electrode plate for use in an electrochemical battery comprises a casting, a flat portion, a perimeter portion and a busbar extending along the perimeter portion. The flat portion and the bus bar are integral parts of the above-mentioned casting. The electrode plate is made of lead, if used in a lead-acid battery, or of another electrode material capable of being formed by casting, if used in other types of batteries. More specifically, the electrode plate has a rectangular shape and a side with a higher current density, and the aforementioned busbar extends along three adjacent sides of the rectangular shape including the side with the higher current density.

Inventors: Swan; David H. (Tatamagouche, Calif.)

Applicant: NameCityState Country Type

Swan; David H. Tatamagush n/a California

Family ID: 38256671

Application No.: 13/385,343

Filed: 15 February 2012.

U.S. Patent8 ,503,162

Seymour 6 August 2013.

Electrode, associated material, method of manufacture and its use

Annotation

The electrode material is created by forming a thin conformal metal oxide coating on a highly porous carbon metastructure. The highly porous carbon metastructure plays a role in synthesising the oxide coating and in providing a three-dimensional, electrically conductive substrate supporting the thin metal oxide coating. The metal oxide includes one or more metal oxides. An electrode material, the process for

The manufacture of said electrode material, an electrochemical capacitor and an electrochemical secondary (rechargeable) battery using said electrode material is disclosed.

Inventors: Seymour; Fraser W. (Middletown, MD)

Applicant: NameCityState Country Type

Seymour; Fraser W. Middletown, MD, USA

Family ID: 44912061

Application No.: 13/189,802

Filed: 25 July 2011.

U.S. Patent 8,404,375

Gaben 26 March 2013.

An electric battery comprising flexible generating cells and a system for mechanical and thermal conditioning of said cells

Annotation

The electric battery includes a plurality of electrical energy generating cells formed by at least one electrochemical cell, which is packaged in a flexible sealed shell, and a system for mechanically and thermally conditioning the cells. The conditioning system forms a structural housing made of a thermally conductive material. The casing has two longitudinal portions and a plurality of transverse portions connecting the longitudinal portions so as to form between the transverse portions of the casing, wherein the generating element is respectively housed. The housing includes a circulation path for the heat transfer fluid. The path has two channels, respectively upstream and downstream, which are formed respectively in the longitudinal member and passages formed in each of the transverse members. The passageways communicate on each side with the upstream channel and the downstream channel.

Inventors: Gabin; Fabien (Courbevoie, FR)

Applicant: NameCityState Country Type

Gabin; Fabien Courbevoie N/A FR

Recipient: Dau Cocam France SAS (Massy, FR)

Family ID: 39868192

Application No.: 12/420,162

Filed: 8 April 2009.

THE FOLLOWING PATENT REPRESENTS A PRE-SELECTED DESIGN OF SIMILAR

TECHNOLOGY :

US Patent8,354,195

Brandstetter et al. 15 January 2013.

Electric storage fuel cell system and method

Annotation

The invention discloses an innovative device and method for electrochemical energy conversion of the fuel cell type with potentially high power density, suitable in particular for autonomous applications such as vehicles powered by electricity The inventive device, called "fuel cell", can be considered as a converted rechargeable battery comprising plates, electrolyte and separators, which can be similar to a conventional known lead-acid battery. Thus, it can be charged from a power source or discharged through an external load with the typically high surge currents typical of rechargeable batteries. However, the main object of the present invention is to provide an innovative additional charging mode for the rechargeable fuel cell by chemically reacting each electrode with fuel and oxidant supplied at specific flow ratios. While all types of prior art fuel cells exchange ions between electrodes via an electrolyte, each of the electrodes of the present fuel cell exchanges ions independently using only its adjacent electrolyte.

Inventors: Brandstetter; Aharon (Kfar-Vradim, IL), Brandstetter; Haim (Bnei-Beraq, IL)

Applicant: NameCityStateCountryType

Brandstetter; Aaron Kfar-VradimN/AIL

Brandstetter; HaimBnei-BerakN/AIL

Family ID: 40379104

Application No.: 12/344,964

Filed: 29 December 2008.

MIX
Papier aus verantwortungsvollen Quellen
Paper from responsible sources
FSC® C105338

Printed by Books on Demand GmbH, Norderstedt / Germany